My World of Science

HEAVY AND LIGHT

Angela Royston

www.heinemann.co.uk/library
Visit our website to find out more information about **Heinemann Library** books.

To order:
☎ Phone 44 (0) 1865 888066
Send a fax to 44 (0) 1865 314091
Visit the Heinemann Bookshop at www.heinemann.co.uk/library to browse our catalogue and order online.

First published in Great Britain by Heinemann Library, Halley Court, Jordan Hill, Oxford OX2 8EJ, part of Harcourt Education.

Editorial: Andrew Farrow and Dan Nunn
Design: Jo Hinton-Malivoire and
Tinstar Design Limited (www.tinstar.co.uk)
Picture Research: Maria Joannou and Sally Smith
Production: Viv Hichens

Originated by Blenheim Colour Ltd
Printed and bound in China by
South China Printing Company

ISBN 0 431 13736 6
07 06 05 04 03
10 9 8 7 6 5 4 3 2 1

British Library Cataloguing in Publication Data
Royston, Angela
Heavy and light. – (My world of science)
1. Weight (Physics) – Juvenile literature
I. Title
531.1'4

A full catalogue record for this book is available from the British Library.

Acknowledgements
The publishers would like to thank the following for permission to reproduce photographs: Alamy Images p. **23**; Collections p. **22**; Corbis pp. **21**, **24**; Fortean Picture Library p. **28**; Photodisc p. **20**; Pictor p. **29**; Robert Harding p. **27**; Trevor Clifford pp. **4**, **5**, **6**, **7**, **8**, **9**, **10**, **11**, **12**, **13**, **14**, **15**, **16**, **17**, **18**, **19**, **25**; Trip/H. Rogers p. **26**.

Cover photograph reproduced with permission of Trevor Clifford.

Every effort has been made to contact copyright holders of any material reproduced in this book. Any omissions will be rectified in subsequent printings if notice is given to the publishers.

Contents

Any words appearing in the text in bold, **like this**, are explained in the Glossary.

Heavy and light

Some things are heavy and some things are light. Both the television and the plant pot are heavy. People find heavy things difficult to lift up.

The ball and the box are light.
Light things are easy to pick up.
The ball of cotton wool is the
lightest thing in the picture.

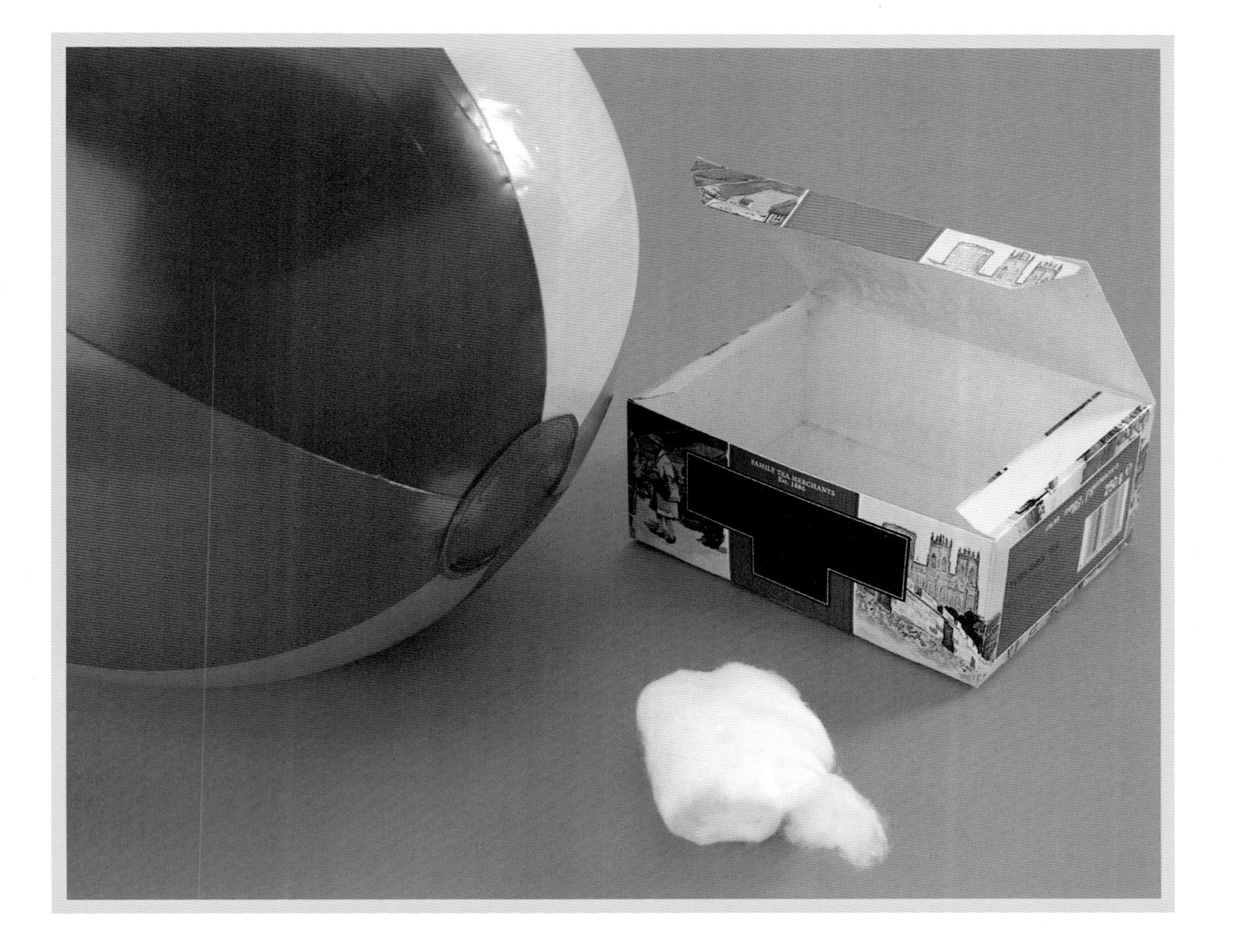

Comparing weights

This girl is holding a book in one hand and a teddy bear in the other hand. She can feel that the book is heavier than the teddy bear.

How heavy something is is called its weight. An orange and an apple weigh about the same. It is hard to tell which one feels heavier.

Weighing

This girl is using **scales** to find out how much the spade weighs. She puts the spade on one side and then adds weights to the other side.

She adds weights until the scales are **level**. The scales are level because the weights are **balancing** the weight of the spade.

Heavy materials

Some **materials** are heavier than other materials. The feather and the pencil are similar sizes. But the pencil is heavier because it is made of wood.

Big things are usually heavier than small things. A paperclip and a car are both made of **steel**. But the car is much heavier than the paperclip.

Light materials

Some materials are lighter than others. This stepladder is easy to lift because it is made of **aluminium**. An **iron** stepladder would be too heavy to lift.

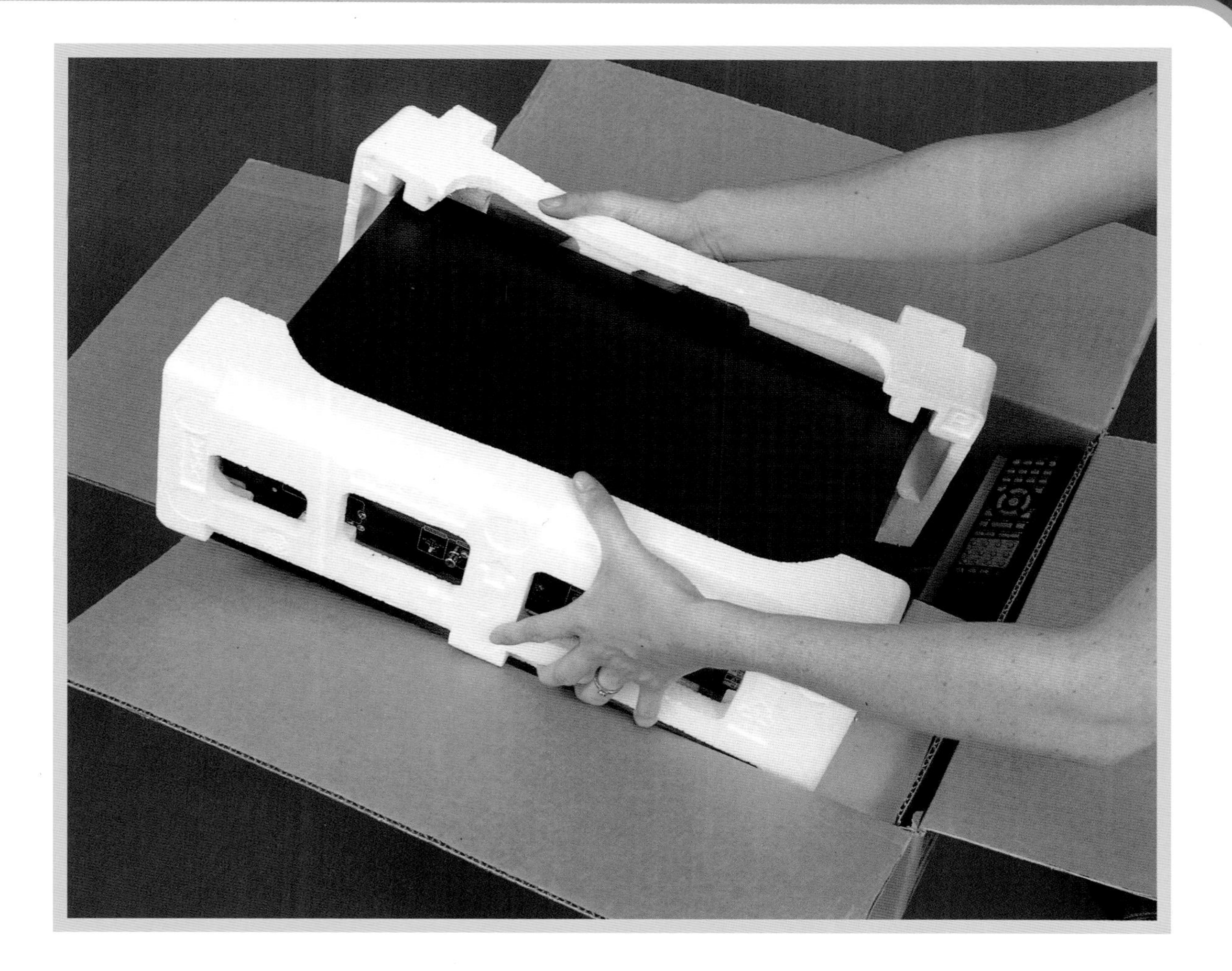

Plastic is a light material. These white blocks are made of a kind of plastic. They protect the video recorder, and they are not heavy to lift.

Water can be heavy

Liquids can be heavy too. This **container** feels light when it is empty. But it is heavy when it is full of water.

A woollen **jersey** is usually quite light. When it is washed, the wool soaks up a lot of water. Then the jersey becomes very heavy.

Filled with air

This balloon is filled with air. It is so light that the boy can easily keep it in the air. He just has to pat it to stop it falling.

These things are all light for their size. This is because they have air trapped inside them. If you squeeze them, they lose some of the air and become smaller.

Does it float or sink?

Things that float in water are light for their size. The duck, balloon and sponge are light because they are all filled with air.

These things cannot float in water because they are heavy for their size. How many things have sunk to the bottom of the tank? (Answer on page 31.)

Heavy things can float

Ships are made of metal and are very heavy. Ships float because they are filled with air. The air makes them light for their size.

This is the top part of a big ship. It hit some rocks and filled with water. This made the ship too heavy to float, and it sank.

Using floats

People use **armbands** when they learn how to swim. The armbands are filled with air. They help people to float more easily in the water.

Lane dividers separate the different parts of this pool. The dividers have floats on them. What keeps the dividers from sinking?
(Answer on page 31.)

Using weights that sink

An **anchor** is made of heavy metal.
It sinks to the bottom of the sea
and digs into the sand or stones.
It stops a boat floating away.

This water plant has a small metal weight attached to it. The weight stops the plant floating around in the fish tank.

Blowing in the wind

Very light things are easily blown by the wind. These autumn leaves are dry and light. Soon the wind will blow them from the trees and lift them through the air.

Dandelion seeds are very light. They can float very far before they sink to the ground. Then new dandelion plants grow where the seeds land.

Lighter than air

This is a huge **airship**. It is filled with a gas called **helium**. Helium is much lighter than air, so the airship floats.

These balloons are also filled with helium gas. They are floating in the air. What will happen if the girl lets go of the strings? (Answer on page 31.)

Glossary

airship an aircraft that is kept in the air by a huge balloon
aluminium a kind of metal that is very light
anchor a piece of metal tied to the end of a long rope or cable. The weight of the metal stops a boat floating away.
armband a band of plastic that can be filled with air. It goes on your upper arm to help you float.
balancing making two things weigh the same
container something that holds something else
helium a gas that is lighter than air and is often used in balloons
iron a kind of metal that is very heavy
jersey a jumper that is made of knitted wool or other fabric
level at the same height as something else
materials stuff that things are made of
scales a tool for weighing things
steel a kind of metal

Answers

page 19

Five things have sunk to the bottom of the tank.

page 23

The floats stop the dividers from sinking.

page 29

The balloons will rise higher and higher into the sky.

Index

Titles in the *My World of Science* series include:

Hardback 0 431 13700 5

Hardback 0 431 13715 3

Hardback 0 431 13712 9

Hardback 0 431 13704 8

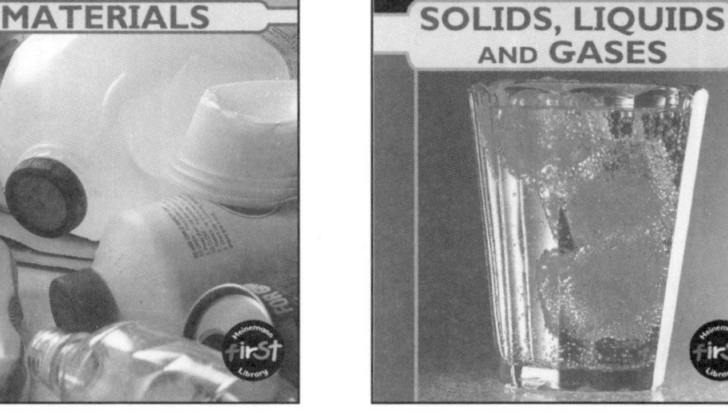

Hardback 0 431 13701 3

Hardback 0 431 13702 1

Hardback 0 431 13714 5

Hardback 0 431 13716 1

Hardback 0 431 13703 X